India's Maritime Heritage

Rear Admiral K. Sridharan, AVSM, (Retd)

CBS

CBS PUBLISHERS & DISTRIBUTORS
NEW DELHI • BANGALORE

First Edition : 2003

ISBN : 81-239-0937-3

Production Director : Vinod K.Jain

Published by:
Satish Kumar Jain for CBS Publishers & Distributors,
4596/1, 11 Daryaganj, New Delhi - 110 002 (India)
E-mail : cbspubs@del.3.vsnl.net.in Web : cbspd.com

Bangalore Branch:
Seema House, 2975, 17th Cross, R.K. Road,
Banasankari 2nd Stage, Bangalore - 560 070
Fax : 080-6771680 E-mail : cbsbng@vsnl.net

Designed by : Smt. Achala and Sri Shivaraj Patil

Printed at : Daksha Printing (P) Ltd., New Delhi - 110 002.

CONTENTS

The Indian Navy to keep in touch with the old art of
sailing, commissioned this modern sail boat.

The emblems of the Indian Navy ships shown in these pages reveal India's heritage over the past several millennia.

FOREWORD

'India's Maritime Heritage' authored by Rear Admiral K Sridharan, AVSM (Retd) is an excellent collage of our rich maritime past and brings out the immense influence it has had on our country in days gone by.

We have now entered a new era where affairs 'maritime' are once again back into focus with the 21st century being recognised as the Century of the Seas.

Admiral Sridharan's book therefore comes at an opportune moment and provides a timely flashback into India's maritime history.

Naval Headquarters
New Delhi

(Sushil Kumar)
Admiral

PREFACE

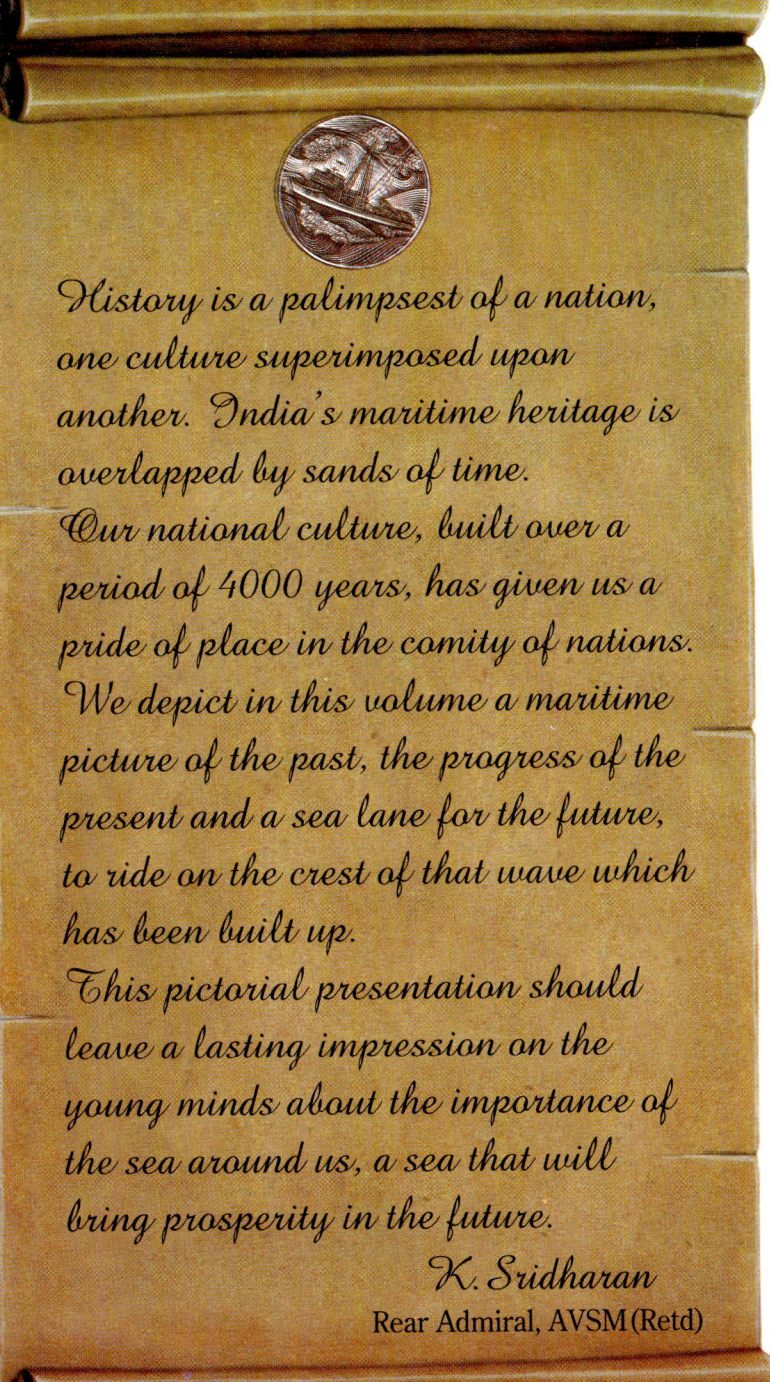

History is a palimpsest of a nation, one culture superimposed upon another. India's maritime heritage is overlapped by sands of time.

Our national culture, built over a period of 4000 years, has given us a pride of place in the comity of nations. We depict in this volume a maritime picture of the past, the progress of the present and a sea lane for the future, to ride on the crest of that wave which has been built up.

This pictorial presentation should leave a lasting impression on the young minds about the importance of the sea around us, a sea that will bring prosperity in the future.

K. Sridharan
Rear Admiral, AVSM (Retd)

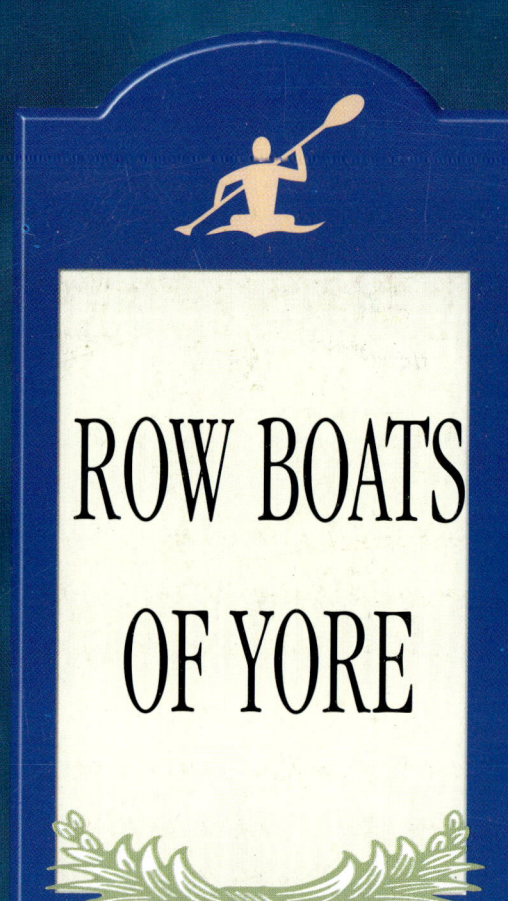

ROW BOATS OF YORE

Man's first adventure afloat.
Harappan was our first sailor

Harappa Culture in 3000 BC was so advanced that the Harappans had ventured out to sea with dug-outs scooped out of tree-trunks, propelled by palm-leaf paddles. The Rann of Kutch was then navigable and the boats evidently went out hugging the coast.

In recent times a discovery of a dock at Lothal at the Little Rann of Kutch, throws more light. The trapezoid shaped structure must have been used either as a boat-pen or a wet basin. It had a provision for regulating the flow of water at low and high tides by means of a spill channel. It was a Harappan engineering achievement way back in 2500 BC.

In some parts of India, in lakes particularly, a boat made of wicker as a bowl shaped basket, rendered water-proof was in use. This type of boat is still being used in lakes.

A houseboat of Kerala, known as "kettuvallom"

South Indians made a humble beginning of using *catamaran* to explore off-shore areas. The English word *catamaran* is a derivative of the Tamil word *Katna maram*, literally meaning *tied wood*. Fishermen of today continue to use catamaran for fishing as the picture shows.

In shallow waters, where there were too many weeds and underwater growth, boatmen were hampered in their rowing with paddle. They adopted an ingenious method of propulsion in such waters by means of a long pole for forward movement. This was prevalent in backwaters of Kerala in south west of India. Even today they follow this tradition in propelling the boat called: *Kettuvallom*.

The idea of erecting a passenger cabin on multi-oared boats was improved upon for Kings to sail in state on the rivers and coastal waters of their Kingdom. The royal barge depicted at the Jagannath Temple in Puri, Orissa, shows a striking feature of a steering oar at the stern which is an advancement.

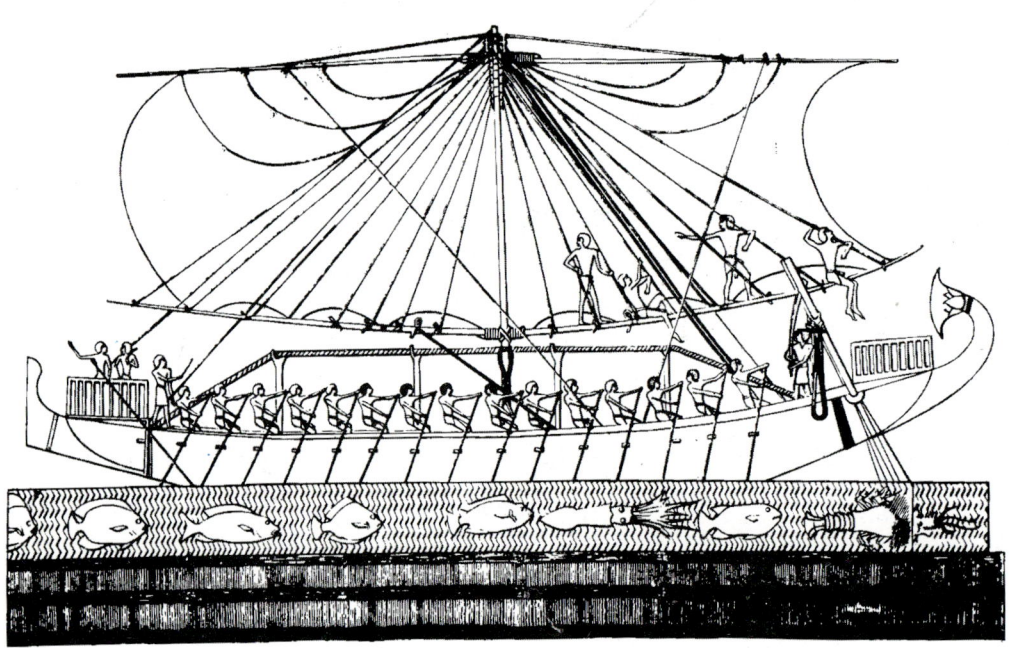

Alexander the Great, during the preparation of his march back from India (327 - 326 BC) came to know from Indians that there was a fairly straight sea route to Mesopotamia. Admiral Niarchus was given orders by Alexander to sail from the mouth of the Indus to the Euphrates. The Admiral acquired a flotilla of river boats some of which were 30-oared as displayed here, built in Punjab by a tribe known as Ksatri/Xethroi. The multi-oared galleys voyaged through 1500 miles, and reached Ormuz, despite gales and rough sea.

This picture shows the Miracle Panel of Stupa at the East Gate front pillar at Sanchi. The boat sculptured shows up-turned stern and a low bow. It carries a dignitary with devotees at the oars. Other devotees standing below are praying for a safe journey.

During the period when multi-oared boats were in use, river battles were waged.

In the bottom panel is a depiction of a multi-oared boat being prevented from landing ashore by a valiant fighter with sword in hand. The hero of this battle is eulogized on the top panel. Such *Hero Stones* can be seen in the north-west coast of India.

The art of rowing multi-oared boats has been kept alive even today in Kerala as a water sport. The picture shows the oarsmen manoeuvring the 100-oared *Chundan*, of 130 feet long snake-boats, to the start line at Alleppey backwaters.

In Kerala conveyance of passengers and farm produce is largely executed by boats - the cheapest mode of travel on water. The picture shows the boats moored at sunset in Kochi backwaters.

BOORA

BUDGAROO

There were several types of boats in use in the Bengal region. From single oared boats, the next evolution was construction of multi-oared boats. The *Boora*, with several oars, which called for the boat to withstand the weight of the oarsmen and keep afloat, was indeed a significant step forward. There was further advancement in the *Budgaroo* and *Oloako* where a cabin had been erected to carry passengers. Finally a multi-sail boat with multi-oars mounting cannon is shown which is the Man of War Prow.

OLOAKO

MAN OF WAR PROW

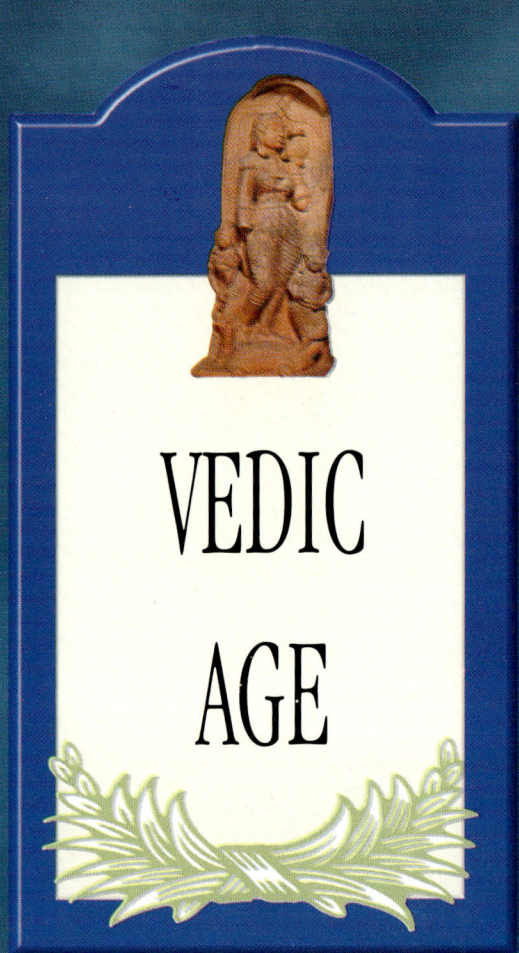

VEDIC AGE

Human devotion to the Cosmic Ocean

India's earliest maritime activity recorded during the Vedic Age (2000 BC to 500 BC) is to be found in Rg Veda, consisting of some 1000 hymns. Several references to the ocean, boats, sea voyages as well as invocations to Varuna, the presiding deity, have been preserved in the Rg Veda and handed down the generations as India's heritage. Varuna, the Lord of the sea, with consort is beautifully depicted in this sculpture.

Indeed some Russian ethnologists opine that Russia's Ded Moroz is Varuna.

From the Vedic Age India's Sea God has been Lord *Varuna* and an invocation to him in the Vedas has since been adopted by the Indian Navy for its Emblem, with the Motto: *Shanno Varunaha* in Sanskrit, meaning: *Be auspicious unto us Oh Varuna.*

On attaining Independence, the Royal Indian Navy's Emblem was re-designed, replacing the Royal Crown with the Ashoka Lion Motif. The old and the new Crests are shown here.

Dwarka at the entrance to the Gulf of Kutch is of legendary fame, the city of Lord Krishna, during Harappa culture which was submerged in later times. Recent excavations reveal what was believed as mythology is indeed factual.

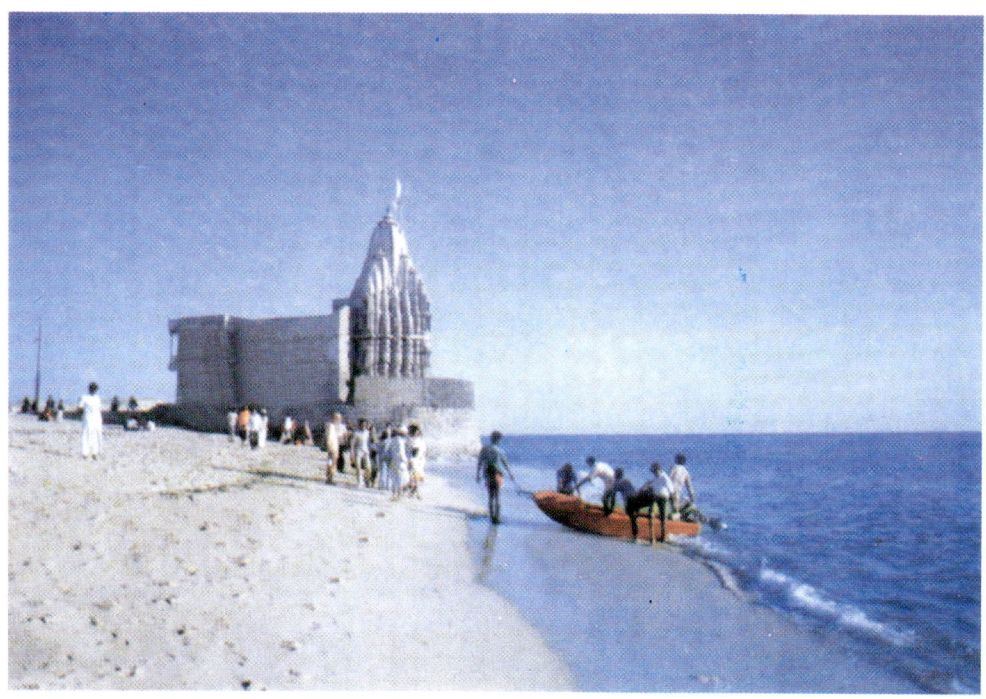

The Dwarkadhish Temple, also known as Varuna Devata Temple, with the ocean lapping its entrance and the treasures underwater dating back to 3000 BC, have been brought to light.

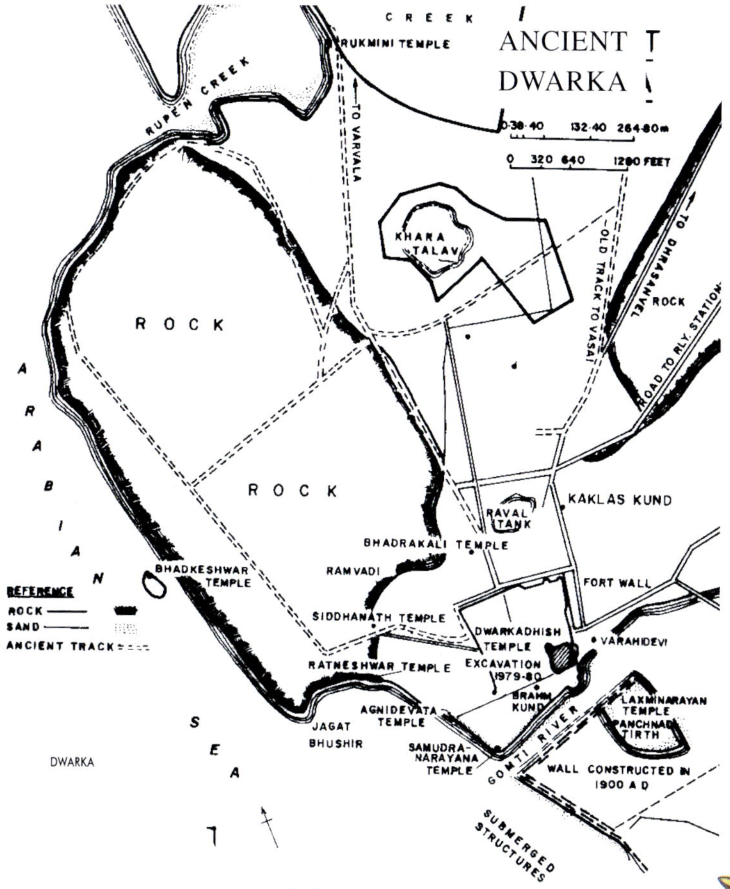
ANCIENT
DWARKA
CREEK
RUKMINI TEMPLE
RUPEN CREEK
TO VARVALA
0 38·40 132·40 264·80m
0 320 640 1280 FEET
KHARA TALAV
OLD TRACK TO VASAI
TO DHRASANVEL
ROCK
ROAD TO RLY. STATION
ROCK
ROCK
ARABIAN SEA
KAKLAS KUND
RAVAL TANK
BHADRAKALI TEMPLE
BHADKESHWAR TEMPLE
RAMVADI
FORT WALL
REFERENCE
ROCK
SAND
ANCIENT TRACK
SIDDHANATH TEMPLE
DWARKADHISH TEMPLE
VARAHIDEVI
RATNESHWAR TEMPLE
EXCAVATION 1979-80
BRAHM KUND
LAXMINARAYAN TEMPLE
PANCHNAD TIRTH
AGNIDEVATA TEMPLE
JAGAT
BHUSHIR
SAMUDRA-NARAYANA TEMPLE
GOMTI RIVER
WALL CONSTRUCTED IN 1900 A D
DWARKA
SUBMERGED STRUCTURES

Indians always deified the ocean, the cosmic ocean on which *Seshasayi Vishnu* reclined when *Brahma*, the creator was thought to have been born out of the lotus supported from the umbilical cord. Hindu mythology depicts God *Vishnu*, reclining in the posture of *Ananthasayanam* on the serpent *Sesha* floating on the ocean , with his hood spread as an umbrella protecting the God's *siras*, the head.

Hindu mythology rightly believed that the ocean is the benefactor offering humanity with the wherewithal for existence. It is indeed so even today, with the riches of the unfathomed depths stored for the benefit of mankind. The legendary depiction in *Samudra Manthanam*, the Vedic reference to the churning of the ocean by means of winding the serpent *Vasuki*, as a rope, around the mount *Mandara*, by both Gods and Demons, to obtain *Amrita*, the nector of God; the ocean that gave jewels churned out of it -a metophoric way to educate the masses of the treasures in the sea.

Some major rivers in India are named after Goddesses, such as the Ganga, Yamuna and Kaveri. Goddesses known as *Nadi Devata* are traditionally represented carrying river water in an urn, considered as *Amruta* (Nectar). Mythology of such depiction has an underlying emphasis. *River is the life blood of human existence, so revere the rivers.* This is the lesson of our heritage, but we are forgetting this tenet and today environmentalists have to remind us not to pollute the rivers.

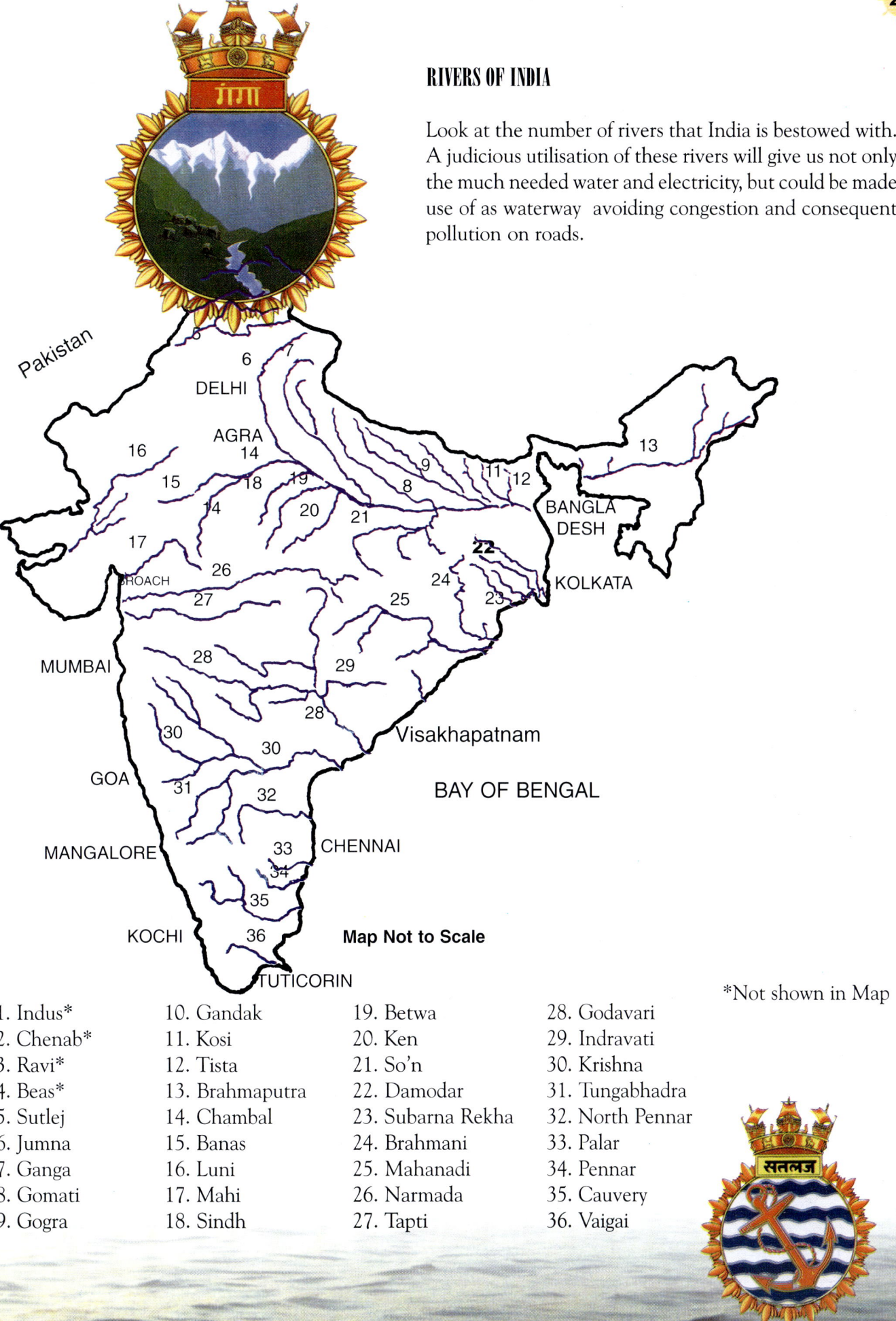

RIVERS OF INDIA

Look at the number of rivers that India is bestowed with. A judicious utilisation of these rivers will give us not only the much needed water and electricity, but could be made use of as waterway avoiding congestion and consequent pollution on roads.

Pakistan

DELHI

AGRA

BANGLA DESH

KOLKATA

MUMBAI

GOA

Visakhapatnam

BAY OF BENGAL

MANGALORE

CHENNAI

Map Not to Scale

KOCHI

TUTICORIN

*Not shown in Map

1. Indus*	10. Gandak	19. Betwa	28. Godavari
2. Chenab*	11. Kosi	20. Ken	29. Indravati
3. Ravi*	12. Tista	21. So'n	30. Krishna
4. Beas*	13. Brahmaputra	22. Damodar	31. Tungabhadra
5. Sutlej	14. Chambal	23. Subarna Rekha	32. North Pennar
6. Jumna	15. Banas	24. Brahmani	33. Palar
7. Ganga	16. Luni	25. Mahanadi	34. Pennar
8. Gomati	17. Mahi	26. Narmada	35. Cauvery
9. Gogra	18. Sindh	27. Tapti	36. Vaigai

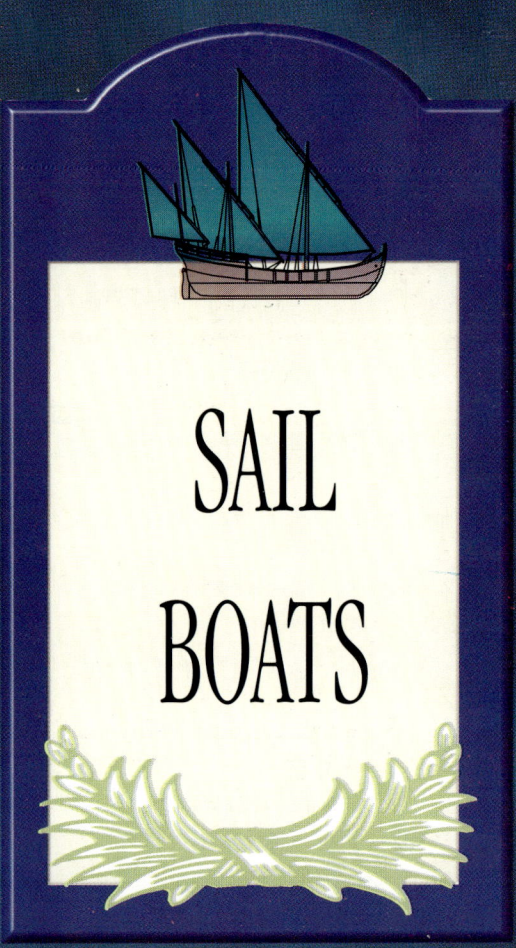

SAIL

BOATS

*Seamen spread their sails
and saw the world*

When the Harappans discovered how to utilise the sea breeze,
they intoduced sails to propel their boats.

In the Eastern waters there was an improved version of catamaran, which was made up of two boats tied together, propelled by oars, as well as sails.

It will be interesting to mention here that Tamil word: *Kapal* meaning ship is being used in Malay language. There is a rocky island in the eastern waters off Malaysia, known as *Alaithangi*, meaning in Tamil *Sea Wave Breaker*. A study of the etymology of marine words, provides supporting evidence for maritime heritage of a country.

Although sails were introduced at this stage of evolution of boat-building, oars were retained as a stand-by when the wind dropped. To negotiate into several coves and creeks off Indian shores, oars were required. The multi-oared galleys with a single mast were, therfore, commonly found. Then decorative bow and stern, carrying at times the emblem of the fief and also his flag at the mast, became an accepted practice.

The next improvement in boat building was the introduction of twin sails. There is numismatic evidence of boats with twin sails in early Andhra period, around 2nd Century AD. A coin minted during the reign of King Yajna Satkarni about 200 AD, depicts a two-masted boat. The two masted boat design has been adopted by the Indian Navy for the Crest of one of its training establishments, situated in Andhra Pradesh, at the suggestion of the author, and named: *Indian Naval Ship Satvahan*, Satvahan being the dynasty of Satkarni.

The discovery of using sea breeze to propel boats brought in sail boats as pictured earlier. Subsequently, improved lateen triangular sails were introduced in dhows which sailed regularly between Indian and Arabian ports. Boats with single sail were further improved by adding more sails.

The Indians also started building larger boats to expand their sea trade. Here is a typical Indian boat with multi-sails which was designed to withstand sailing in the high seas at all seasons.

The next improvement was to increase the size of boats to be able to carry heavier loads and to be propelled by multi-oared and multi-masted vessels, that India largely emulated from the Portuguese galleys which started frequenting Indian ports.

Deifying of an Ocean-God is not peculiar to India. Even in the early days, the Greek God of the Sea, *Poseidon*, identified with the Roman Neptune, was so much revered and it is surmised that in one of the voyages from Europe to India, this idol, created by Lissipus, was found installed when excavations were undertaken in Indian territory at Brahmapuri, Kolhapur in Maharashtra region.

Invoking the blessings of God or Goddess by sailors is universal in concept. Influence of the Portuguese culture in Goa is evident in this icon of Mary with Jesus in hand, standing on a ship pedestal.

Chinese junks had voyaged all the way to Kerala as early as 125 A.D. for India's pepper and ginger. In return they traded with Chinese jars and paper. The Chinese fishing nets in Cochin (Kochi) are a direct evidence of their interaction. The clever contraption is in use even today. Chinese influence is also noticed on the east coast of India. In Nagapattinam (Negapatam), a port in Pallava/Chola period, the Chinese constructed about 11th century A.D., a Buddhist temple, in brick, which came to be known as "China Pagoda", where a bronze statue of the Buddha was installed.

The Portuguese galleons were large sized boats built to withstand turbulent high seas during their long voyages via the Cape of Good Hope.

Pearl and coral fishing was India's age old traditional way of exploiting the rich reserves of the sea. Pandyan pearls became so famous that foreigners were constantly looking for Indian pearls.

The Indian Ocean was famous also for its corals (top) and shells (bottom) even in the historic past. Indian corals are sought after so much for their excellent hue, variety and sturdiness; hence extensively used for making ornaments.

Here we depict a variety of sailing ships which were plying in the Indian seas during this period.

As the interaction with other countries developed , boat designs were modified from time to time.

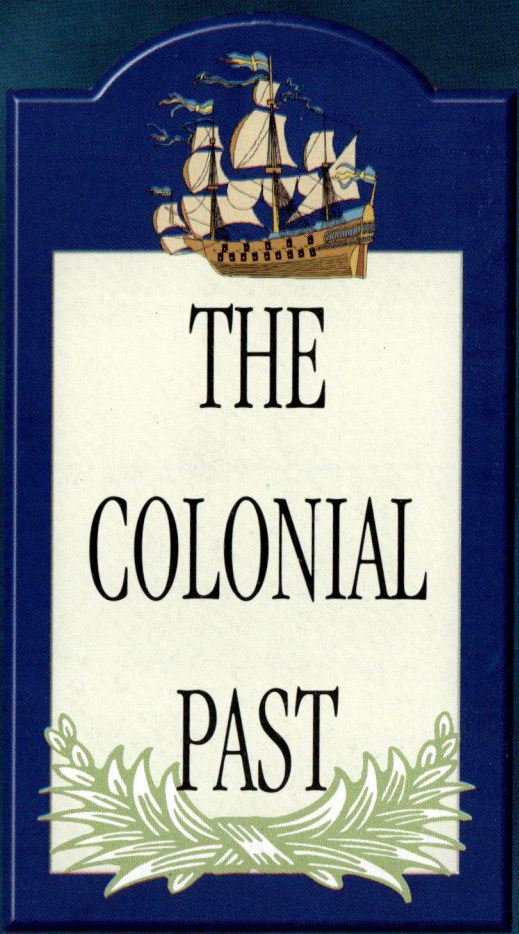

THE COLONIAL PAST

The "silk route" overland through Khyber Pass was found to be more risky and expensive due to the terrain and the levy *en route* by locals, when compared to the semi-sea route to India via the Euphrates and the Persian Gulf, which came into vogue from 7th century onwards. The all-sea route known as the "Cape Route" was established in the 15th century. After the opening of the Suez, the "Canal Route" came into existence from the 19th century. The sea route from the west coast of Africa was also known to the Arab seamen from 8th century onwards. The Indians established as early as 1st century, a regular sea route from Puri to Java via Rangoon and the Malacca Straits.

The fight for sea power in the Indian Ocean

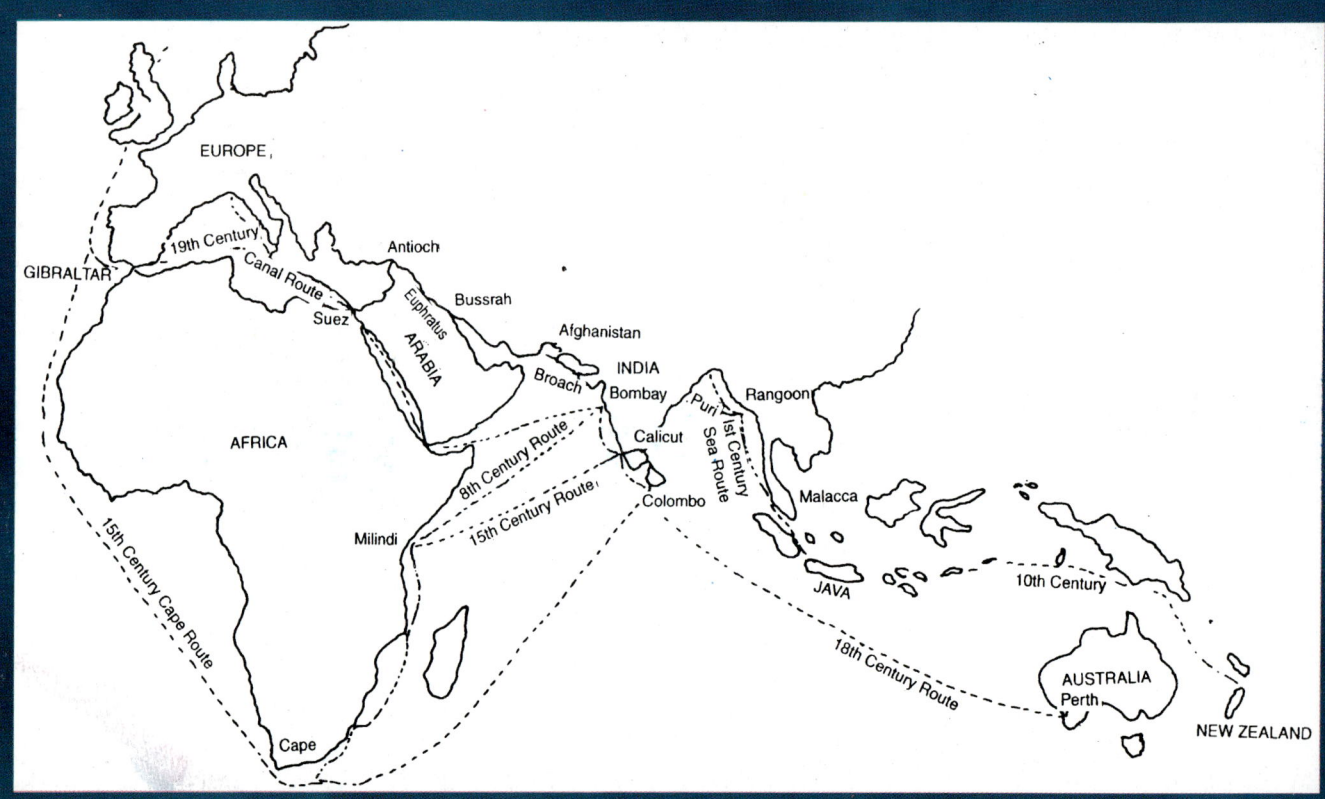

The quiet fishing village inhabited by Koli tribes in the cluster of seven islands, where Mumba Devi was worshipped, slowly but surely changed over the centuries to become world famous as the city of Bombay.

With the entry of the Portuguese, the Castle abutting the Arabian Sea became a landmark. The gate of the entrance to the "Castle Barracks", as it is now known, is flanked on top by two pantalooned Portuguese soldiers carved in stone on massive wooden blocks, one of which is pictured here.

The other impressive Portuguese contribution is the Sun Dial which stands ten feet high with carvings of busts of men, monsters and mammals put together in bizarre form.

Having somewhat established themselves in Surat, the British looked for a better haven on the east coast of India. In 1639, they got the lease of Madras (present Chennai) from the Raja of Chandragiri and fortified the sea port, naming it Fort St George.

Here the Madras roadstead is seen where the British are embarking and disembarking. It is interesting to see the ladies being carried ashore to save them wading through the sandy beach.

The British transformed the scenario of Bombay.
Here is the starting of the brick and mortar intrusion on nature's beauty - the Port Trust and the Customs House.

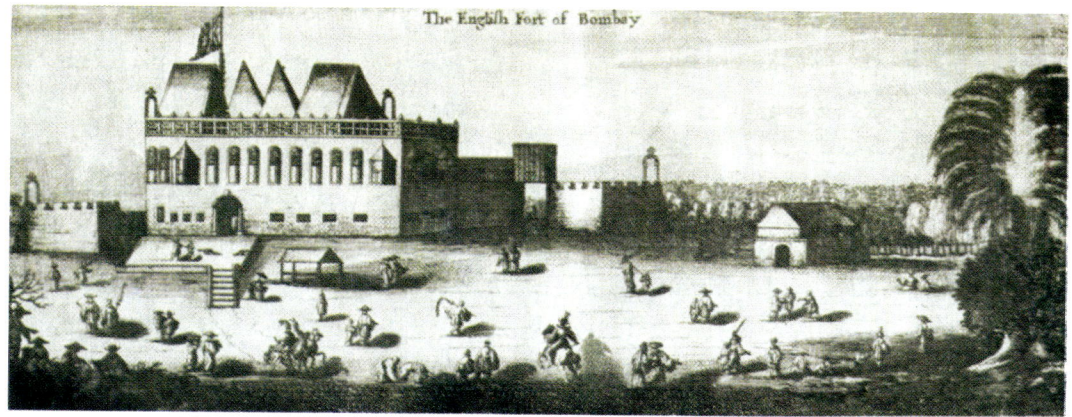

In 1668, the East India Company obtained Bombay (present Mumbai) by a stroke of good fortune on nominal rent from Charles II, who secured the island through his marriage to Catherine of Braganza. By 1687, Bombay became the main British entrepot. The English fort of Bombay is seen in this picture.

The Maratha naval power challenged the British occupation of Bombay from the island of Khanderi, where Shivaji constructed a fort in 1679. Several attempts by the British to conquer the island failed till 1750. The Khanderi Fort, situated 11 miles south of Bombay, has been pictured here from the air.

Here is yet another island fort at Janjira, situated 50 miles south of Bombay, which was a stronghold of the Sidis. While the Moghul naval interest was centred at Cambay, at the mouth of the Gulf of Cambay, the Sidi sea-borne commerce was based at Janjira.

The Dutch, primarily due to their command of the sea, were able to establish in the 17th century, a leading role in maritime commerce. The Dutch thrust was more on the east coast of India than the west coast.

The picture shows a lighthouse built by the Dutch in 1867, off the shores of Visakhapatnam on the east coast of India.

Maratha boats attacking British Ships

GANGA

Built during World War II as a "Hunt" class type II escort vessel, launched on 10 March 1941 and commissioned on 16 October 1941 as HMS "Chiddingfold" served with distinction in World War II in the Atlantic escorting North Sea convoys.

Transferred to the Indian Navy in May 1953 and commissioned as INS "Ganga" she formed part of the 22nd Destroyer Squadron and took part in all fleet activities, including the 1971 operations against Pakistan. "Ganga" decommissioned on 31 July 1972.

RANJIT

Built as an "R" class destroyer for the Royal Navy and commissioned as H.M.I.S. "Ranjit" on 4 July 1949. Formed part of the 11th Destroyer Squadron and took part in all fleet activities including the 1971 operations against Pakistan. "Ranjit" was decommissioned on 30 September 1975.

KRISHNA

"Black Swan" type sloop specially built for the then Royal Indian Navy by M/s Yarrow and Co. Ltd., Glasgow and initially commissioned as H.M.I.S. "Krishna" on 23 August 1943. Was manned by the Royal Indian Navy and saw service in the Atlantic and Western Approaches and later in the Far East during World War II. Converted into a cadet training ship in 1955 and was part of the 12th Frigate Squadron until 1972. She continues to serve and is at present a part of the 1st Training Squadron and trains midshipmen of the fleet.

CAUVERY

"Black Swan" class sloop and sister ship of "Krishna", was also built by M/s Yarrow and Co., Glasgow and commissioned on 21 October 1943.

"Cauvery" saw service in the Atlantic and Western Approaches and the Far East in World War II, was one of the first Indian warships to enter Japanese waters and formed part of the Commonwealth occupation forces in Japan. "Cauvery" has taken part in the 1961 Goa operations and the 1971 operations against Pakistan. Converted to the training role in 1972, "Cauvery" formed part of the 1st Training Squadron until she was decommissioned on 30 September 1977.

TIR

Built as a "River" class frigate by Charles Hill and Sons of Bristol and commissioned as H.M.S. "Bann" on 07 May 1943. Transferred to the Royal Indian Navy on 03 December 1945. Saw service in the Far East and the East Indies during World War II. Renamed "Tir" in Feb. 1947 and used as the boys' training ship. Converted into a cadets' training ship in June 1950. "Tir" carried out this role with distinction for 27 years. She has trained all the cadets and midshipmen of the Indian Navy during the last quarter century. "Tir" was decommissioned on 30 September 1977.

KONKAN

An ocean minesweeper of the "Bangor" class. "Konkan" was built by Lobnitz and Co., Renfrew, Scotland in April 1942. Transferred and commissioned into the Royal Indian Navy on 29 Janjuary 1947. "Konkan" did both escort and minesweeping duties in the Atlantic, Mediterranean and Indian waters during World War II. Along with "Bombay", "Bengal" "Madras" "Rohilkhand" and "Rajputana" she formed the 31st Minesweeping Squadron. "Konkan" was converted into a diving tender in 1964 and took part in the 1971 operations against Pakistan. She was decommissioned on 31 July 1972.

DHARINI

Built by M/s Foundation Maritime Ltd. of Pictou, Nova Scotia, Canada, "Dharini" was initially commissioned as "Le Petite Hermione" on 10 July 1944. Transferred to the Indian Navy as a Depot and Repair Ship. She was renamed "Dharini" and recommissioned on 2 April 1952. "Dharini" carried out the role of a logistic support ship until 1959 when she was converted into a depot ship for support of coastal minesweepers. In 1971 "Dharini" was assigned the role of mother ship for the missile boats and provided logistic support to these ships until she was decommissioned on 31 December 1975.

SHAKTI

Specially built for the Indian Navy by M/s Naval Meccanica of Naples, Italy, "Shakti" was the only fleet tanker of the Indian Navy until the arrival of "Deepak" in 1967 and along with "Dharini" comprised the Fleet Replenishment Group of the Navy. "Shakti" was decommissioned on 31 December 1967.

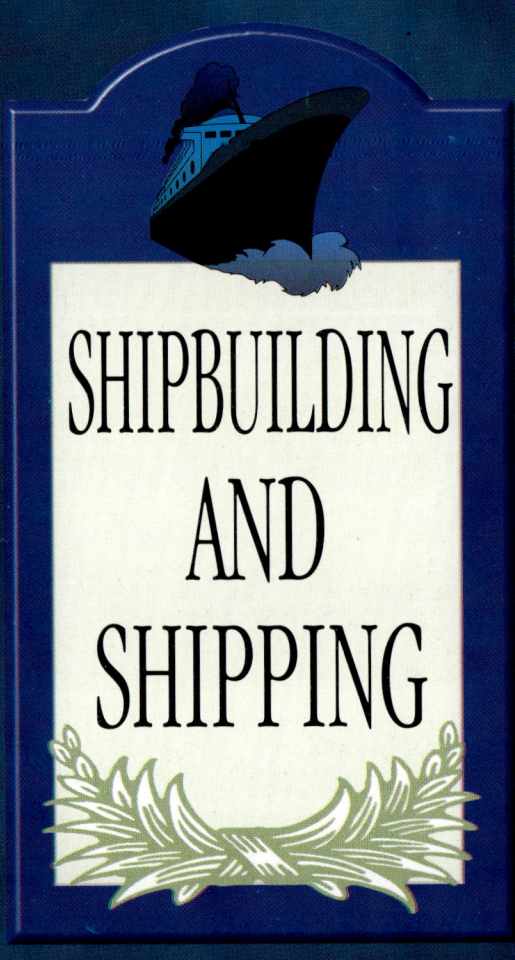

SHIPBUILDING AND SHIPPING

India's salvation for achieving self-sufficiency

The *Trincomalee*

Afterview of the *Trincomalee*

The British initiated building of sail ships at Bombay because they decided to pull out of Surat. They succeeded in their venture and the Wadia Master Builders emerged. The Wadias built as many as 380 sail ships, big and small in the Naval Dockyard, Bombay, from 1736 to 1932. The ships were built out of teak wood which withstood the weather much longer than oak. It is a wonder of the world that the *Trincomalee*, a 46-gun sail ship, (seen in this picture) re-named, *Foudroyant*, launched in 1817 at Bombay is afloat in British waters even today.

The *Foudroyant*

The 84 gun Sailing Ship "ASIA" built in Mumbai Dockyard in 1824.

India's shipbuilding activity declined after the Industrial Revolution, when a sea change from wood to iron and sail to steam took place. It did not revive until the Scindia Steam Navigation Company started a shipbuiding yard at Visakhapatnam on the east coast of India. The picture above shows the first merchant ship, named: "Jalausha", being launched on 14 March, 1948 and the picture below shows the launching of "Jalprabha" on 20 November, 1948. After two years the yard was re-christened: "The Hindustan Shipyard Ltd.", as a public sector undertaking.

REPAIR DRY DOCK

The Hindustan Shipyard Ltd., was modernised as years went by. The picture above depicts the present Repair Dry Dock. The picture below shows outer view of the covered Building Dock.

BUILDING DOCK

Hindustan Shipyard Ltd (HSL) started constructing Offshore Platforms and Drill Ships for Oil & Natural Gas Corporation (ONGC) from 1980. Picture above shows substructure 'Jacket' of Offshore Platform, fabricated by the Yard, being loaded out.

This is "Sagar Bhushan", a Drill Ship for ONGC built by HSL in 1987.

This is "Akbar", which underwent extensive steel renewal/repair at HSL in 1995.

This is "Maharashtra", a 42,750 DWT Bulk Carrier, built by HSL and delivered to the Shipping Corporation of India (SCI), on 6 January, 1996.

This picture is the launching on 11 December, 1996, of a Passenger-cum-Cargo vessel, built by HSL for Andaman & Nicobar Administration.

This is "Goa", a 42,750 DWT Bulk Carrier built by HSL and delivered to SCI on 15 January, 1998.

"Swaraj Dweep", a Passenger-cum-Cargo vessel, built by HSL and delivered to Andaman & Nicobar Administration on 9 December, 1999.

"Tamil Nadu", a 42,750 DWT Bulk Carrier, built by HSL, delivered to SCI on 15 September, 2000.

The Mazagon Dock came into existence at Bombay in 1770. The joint owners, the Peninsular & Oriental Steam Navigation Company and the British India Steam Navigation Company used the Mazagon Dock initially to dry dock their ships. The ship repair facilities were introduced and its 150 feet slipway launched many a famous ship. The Government of India took over the yard in 1960 and started modernising it as Mazagon Dock Ltd.

We show in the succeeding pages, the activities of the Yard and its achievements over the years.

Ritchie Dry Dock

Cargo Vessel, *Asean Progress*. Built by MDL

Major modification done to The *Atlas -1* of the USA

Cutter Suction Dredger, *Meena. Built by MDL*

AL VARO-DE-BAZAN, a Spanish super tanker, repaired by Mazagon Dock

Cutter Suction Dredger, built by Mazagon Dock.

Saudi Arabia's Drill Ship, ARAB DRILL -2, repaired by Mazagon Dock.

SHIPS BUILT BY MAZAGON DOCK LTD

Off shore Patrol Vessel, CGS *Varuna*

Frigate, INS *Taragiri*

Frigate, INS *Gomati*

The Mazagon Dock also constructed Drilling
Rigs and Off-shore Supply Vessels

Sagar Uday, Drilling Rig

Feroze Gandhi, an off-shore Supply Vessel

Sindu-14, Off-shore Supply Vessel

Goa Shipyard

In 1957 the Goa Shipyard was established by the name of "Estaleiros Navais de Goa", when Goa was one of the Portuguese colonies. In December,1961, the Yard became the property of the Government of India, when the Portuguese rule ceased.

Starting with repairs to barges, the Yard, seen in the picture, built small ships and craft, as for example the first barge *Pagatim I* shown here, as well as a modern fishing vessel the *Menaka Dan* .

Pagatim I

Menaka Dan

GRSE

Mahaganga

Skipjack

The Garden Reach Shipbuilders & Engineers Ltd. (GRSE), Calcutta, since 1975 is engaged in the construction of a series of Seaward Defence Boats, Ocean Going Tugs, Hydrographic Survey Vessels, Inshore Patrol Vessels and dredgers.

The launching of the *Mahaganga*, Asia's biggest dredger in the 1970s. and a Research Vessel *Skipjack*, both built by GRSE are shown here.

COCHIN SHIPYARD

Cochin Shipyard was built with Japanese collaboration during the decade 1972 to 1982. This is an aerial view of the Yard.

The Cochin Shipyard built bulk carriers for large shipping concerns of India. The picture above is a bulk carrier built for Chowgule Steamship Co.

Cochin Shipyard has the distinction of building in 1990 this ship, "Motilal Nehru", which is the first Indian built oil tanker of 86,000 DWT for the Shipping Corporation of India.

Cochin Shipyard also built specialised vessels, such as the tug "Cheetah" built for Kandla port.

SS *Loyalty* was the first merchant ship of the first Indian shipping company, the Scindia Steam Navigation Co. Ltd. The ship sailed to the UK on 5th April, 1919. This date is being celebrated as India's Maritime Day from 1964 onwards.

The Shipping Corporation of India, on its formation in 1961, started with just 19 ships, totalling 1.38 gross registered tons (GRT) and achieved a peak figure in 1985 with 158 ships, with a total GRT of 33 tons.

We are showing a range of three ships – a Container Ship, a Cargo-cum-Passenger ship and an Oil Tanker, owned and operated by the Shipping Corporation of India.

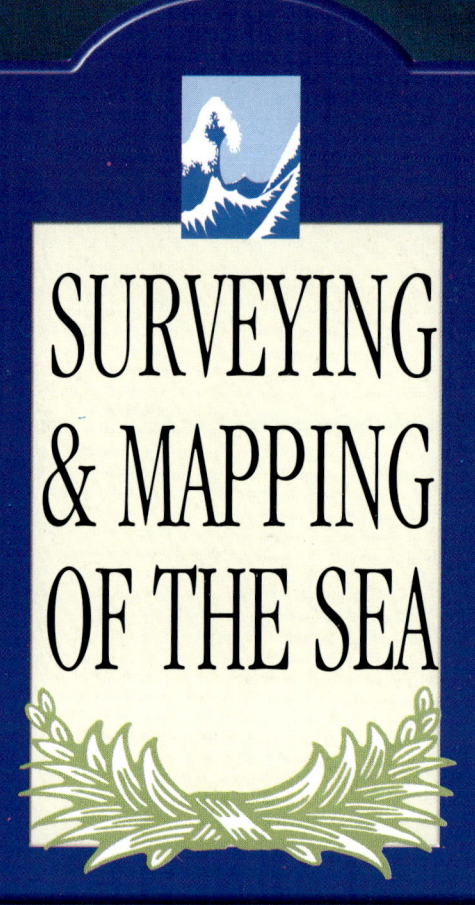

SURVEYING & MAPPING OF THE SEA

Silent Service makes

navigation safe at sea

Surveying and mapping of the sea started in Asian waters shortly after the mapping of the stars. We depict here some ingenious ways of how cartography developed.

An Indian chart of the coast of the Arabian and Red Sea in the 18th century.

Year 1881

Year 1924

The name of *Investigator*, christened for a marine survey ship in 1881, for mapping the sea, has been adopted from generation to generation. A variety of ships which served as *Investigator* through a century is displayed here.

Inset on top is a Pattamar sail boat which was used for marine survey of off-shore sea approaches along the Indian coast line in the 18th century.

Year 1950

Year 1989

MODERN BALANCED FLEET

*Ships take wings & soar
sky high*

*Dive below the sea for
stealth & ocean wealth*

*The Navy rules the waves
in the Missile Age*

THE OLD AND THE NEW I.N.S. DELHI

This is a Leander Class cruiser, built in U.K. in 1933 and commissioned as HMS Achilles. It was loaned to the New Zealand Navy and took part in the Battle of the River Plate in December, 1939.

On attaining independence, this ship was bought by India from the UK, after it was commissioned as HMIS Delhi on 5th July, 1948. It served as the first Flag Ship of the Indian Navy and later as training cruiser. The Delhi was unfortunately sent to the ship breakers as it could not be retained as a museum piece and maintained as India's naval heritage due to financial constraints.

This is a six-inch gun turret from the *Delhi* (Ex-*Achilles*) gifted by India to New Zealand and mounted as a memorial by that country at Auckland.

The name *Delhi* has been retained for the new guided missile destroyer shown in this picture. This is an indigenously built ship, commissioned in 1997.

Ten years after India obtained independence, a second cruiser, the *Nigeria*, was purchased from the UK, refitted with up-dated equipment and commissioned as Indian Naval Ship *Mysore* on 29th December,1957. The *Mysore* took part during the Goa operations of 1961 in taking over Anjadiv island.

The ship was mounted with nine 6" guns on three turrets. In 1985 the ship was decommissioned and sent to the ship breakers.

The Gun Turret of *Mysore*

The *Mysore*

THE INDIAN NAVY ABOVE THE SEA

In 1953 a Naval Air Station was established at Cochin (Kochi). From a humble beginning of acquiring Sealand aircraft and the Firefly for target towing, the first aircraft carrier, the *Vikrant*, was commissioned in 1961 with a number of squadrons of *Alize & Sea Hawk* aircraft. The *Alouette & Seaking* helicopters were also acquired. Then the reconnaisance and anti-submarine aircraft, starting with the Super Constallations were inducted. On 21st December, 1983 the first Indian *Sea Harrier* landed on the flight deck of INS *Vikrant*. The Vikrant has been paid off and is being converted as a museum piece. The second aircraft carrier, the *Virat*, was acquired in 1987.

We show here and on the following pages a series of pictures of the Navy above the sea.

Sealand aircraft flying in formation

A *Firefly* of the Indian Navy

Sealand aircraft touching down at Ernakulam Channel

"*Sea Harrier* flying past the Vikrant"

INS *Viraat*, India's second aircraft carrier

Sea King in operation

Allouette on training operation

The *Islander* aircraft on reconnaissance mission

Helicopter taking off from the Carrier

The successful expedition to the Antarctica was given large logistic support by the Indian Navy. Pilots ferried naval helicopters - the *Chetaks* - to far off South Pole from the base camp. In the 7th Expedition from November, 1987 to March,1988, the *Chetaks* logged 250 hours of total flying on reconnaissance, supply and communication missions.

THE ERA OF ROCKETS AND MISSILES

The acquisition of the *Osa* class rocket boats from the erstwhile USSR in 1970, was the Indian Navy's introduction to the Missile Age. Then the *Nilgiri*, the first Leander Class frigate, built in India in 1972, was fitted with a single *Seacat* guided missile launcher and was followed with two *Seacat* surface-to-air guided missile launchers in the 5th and 6th *Leander* Class frigates built in the 80's.

The success of the Missile Boats during the Pakistan War of 1971, gave the Indian Navy an impetus to replace conventional guns on board ships with missiles. In 1975, INS *Trishul* was retrofitted in India with missiles, replacing the conventional guns.

The Indian designed *Godavari* Class frigate, INS *Ganga*, was equipped with long range surface-to-surface missiles and point defence surface-to-air missiles. From 1988 to 1990, the indigenously built missile corvettes joined the Indian fleet.

India has pointed out to the foreign powers committed to the Missile Technology Control Regime that India's concern to develop its own missile technology is purely a defensive necessity to prevent any external attack.

In this and the following page, we depict the progress of India's Missile Age.

A Missile Boat built in erstwhile USSR

Navy's Missile Boat in action

Missile firing practice at sea

Eastern fleet excercise

THE INDIAN NAVY BELOW THE SEA

On 8 December,1967, India commissioned a conventional submarine INS *Kalvari* , in collaboration with the erstwhile USSR. This was followed by three more submarines, the *Karanj, Khanderi & Kursura* in the next three years under the same collaboration. From 1973 to 1975, India purchased from USSR four more submarines.

On gaining experience below the sea, two more modern SSK (Submarine-Submarine Killer) boats were acquired from West Germany in 1986.

Then two indigenously built submarines, under West German collaboration, came into Service. Unfortunately, this project had to be fore-closed due to suspected leakage of drawing details prejudicing the security of this type of submarine at sea.

India had the unique opportunity to handle a nuclear submarine commissioned as INS *Chakra* , given on loan from the erstwhile USSR for a period of 3 years from 1988 to 1991. This experience will stand in good stead for the Service in future.

We depict some of these submarines in this and the following pages.

SSK Boat, *Shishumar*

Kilo Class Boat entering Cochin harbour

Two submarines of the Kalvari
Class *(Foxtrot)* on joint exercise
at sea.

A Kilo class submarine of the Indian Navy

INS *Shalki*, the first Indian built Submarine launched at Mazagon Dock on 30 September 1989

INS *Chakra*, loaned from erstwhile USSR

Indian Navy Submarine preparing for Review

The HDW Type 209 - 1500 submarine

VIGIL BY THE COAST GUARD

India's Coast Guard Service was started in 1977 in order to safeguard off shore installations, protect maritime environment from pollution, assist Customs Authorities in anti-smuggling operations, enforce Maritime Zone of India Act and last but not least, help ships in distress. Consequent to India obtaining the status of "pioneer investor", involving exclusive right to explore 1,50,000 sq.kms of sea in the Indian Ocean under the UN agreements, the responsibility of effectively protecting the nation's right, has primarily devolved on the Coast Guard.

C.G.S. *Vijaya*, Offshore Patrol Vessel

INDIA'S MAJOR PORTS

- Mumbai
- Nhava Sheva
- Marmugao
- New Mangalore
- Kochi
- Tuticorin
- Chennai
- Visakhapatnam
- Paradeep
- Kolkata
- Kandla
- Ennore

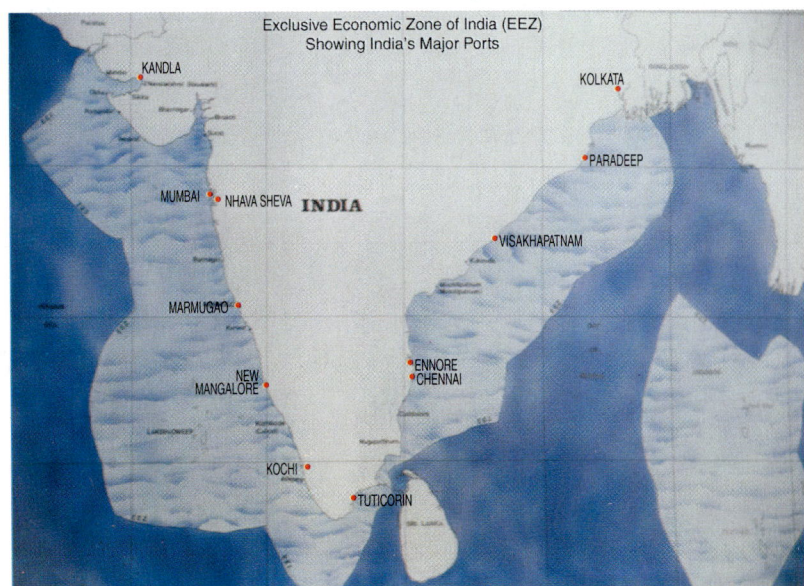

Exclusive Economic Zone of India(EEZ)

Dornier Aircraft

C.G.S. *Vivek,* Offshore Patrol Vessel

BALANCED FLEET OF THE INDIAN NAVY

The Navy on the sea, above the sea, and below the sea

WE AID OUR NEIGHBOURS

In November, 1988, the young Republic of Maldives was facing a coup. India responded with military aid immediately after the President of Maldive Islands appealed for help. Indian Army troops were landed at key locations. The coup leaders high-jacked a merchant ship, the *Progress Light*. The Indian warships, *Godavari & Betwa*, after some firing, captured the ship in the Indian Ocean.

Landing Ship Tank

The *Progress Light,* being straddled

PRESENTATION OF COLOURS
TO THE INDIAN NAVY

Presentation of King's Colours
To The Royal Indian Navy
By H.E.The Right Hon'ble Lord Brabourne, GCIE, MC, Governor of Bombay in 1933

By President Rajendra Prasad
To The Indian Navy on
27th. May, 1951

By President Giani Zail Singh
To The Southern Naval Command on 20th November 1984.

By President Giani Zail Singh
To The Eastern Naval Command on 15th April 1987.

By President R. Venkatraman
To The Western Naval Command on 22nd February 1990.

Acknowledgements

GRATEFUL ACKNOWLEDGEMENT OF SOURCE

1. Naval Headquarters, New Delhi
2. National Museum, New Delhi
3. State Trading Corporation of India
4. National Institute of Oceanography, Goa
5. Ms. Nomita Kamdar
6. Lothal Museum
7. Archaeological Survey of India
8. Sree Padmanabhaswamy Temple, Tiruvanandapuram
9. Commander V.S.P.Mudaliar,I.N.(Retd)
10. The Hindu, Chennai
11. Commander B.Lobo,I.N.(Retd)
12. Prince of Wales Museum, Mumbai
13. National Maritime Museum,Mumbai
14. Director of Archives & Archaeology, Goa
15. Deccan College Post Graduate & Research Institue,Pune
16. Shipping Corporation of India
17. Kolhapur Museum
18. Heras Institute of Indian History & Culture, Mumbai
19. Rear Admiral F.L. Frazer AVSM (Retd.)
20. Goa Shipyard Ltd
21. Mr.Sadashiv Gorakshkar
22. Ms.Kalpana Desai
23. Dr.S.R.Rao
24. Society for Marine Archaelogy, Goa
25. Hugh & Coleen Gantzer
26. Scindia House, Mumbai
27. Cochin Shipyard Ltd.
28. Hindustan Shipyard Ltd.,Visakhapatnam
29. Admiral R.L. Pereira, PVSM, AVSM
30. Vice Admiral R.P.Sawhney,PVSM (Retd)
31. Western Naval Command, Mumbai
32. Eastern Naval Command, Visakhapatnam
33. Southern Naval Command, Kochi
34. Mazagon Dock Ltd,Mumbai
35. Garden Reach Shipbuilders & Engineers Ltd.,Kolkata
36. AdmiralJ.G.Nadkarni,PVSM,NM,AVSM,VSM (Retd)
37. Armed Forces Information Officer,New Delhi
38. Indian Coast Guard Headquarters, New Delhi
39. Naval Public Relations Officer,New Delhi
40. Quarter Deck, Naval Headquarters
41. New Zealand Navy
42. Vice Admiral VEC Barboza, PVSM,AVSM & Bar (Retd)
43. Mr. R.A. Wadia
44. Mr. O.K. Nambiar
45. Admiral L. Ramdas, PVSM,AVSM,Vr.C,VSM (Retd.)
46. Mrs. Anjolie Ela Menon
47. Publications Division, Ministry of I&B, Govt. of India
48. Mr. R.K. Mookerji
49. Admiral Sushil Kumar, PVSM,UYSM,AVSM,NM (Retd)